49 EASY-TO-BUILD ELECTRONIC PROJECTS

No. 1337
$10.95

49 EASY-TO-BUILD ELECTRONIC PROJECTS

BY ROBERT M. BROWN & TOM KNEITEL

TAB BOOKS Inc.
BLUE RIDGE SUMMIT, PA. 17214

FIRST EDITION

FOURTH PRINTING

Printed in the United States of America

Reproduction or publication of the content in any manner, without express permission of the publisher, is prohibited. No liability is assumed with respect to the use of the information herein.

Copyright © 1981 by TAB BOOKS Inc.

Library of Congress Cataloging in Publication Data

Brown, Robert Michael, 1943-
   49 easy-to-build electronics projects.

   "TAB book #1337."
   Includes index.
   1. Electronic apparatus and appliances—Design and construction—Amateurs' manuals. I. Kneitel, Thomas S., joint author. II. Title.
TK9965.B738   621.381   80-28861
ISBN 0-8306-9630-X
ISBN 0-8306-1337-4 (pbk.)

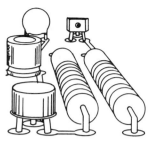

Contents

	Preface	7
1	Wireless Home Broadcaster	10
2	Hand Motion Music Maker	12
3	Personal Code Practice Oscillator	14
4	One-Transistor AM Radio	16
5	All-Purpose Signal Generator	18
6	Echo-Chamber Amplifier	20
7	Automatic Safety Flasher	22
8	Headset-To-Speaker Adapter	24
9	Batteryless Transistor Receiver	26
10	Magic Meter Sounder	28
11	Amazing Electronic Music Maker	30
12	Miniature FM Radio	32
13	CB Field Strength Meter	34
14	Dynamic Mike	36
15	Handy Signal Tracer	38
16	Legal CW (Code) Radio Transmitter	40
17	High-Power AM Receiver	42
18	80 to 150 MHz VHF Receiver	44
19	Supersonic Eavesdropper	46
20	Powerful Code Practice Monitor	48
21	Automatic AC Drive Controller	50
22	Efficient 2-Transistor Audio Preamplifier	52
23	Automatic Auto Light Reminder	54
24	Electronic Moisture/Rain Alarm	56

25	Light Beam Communications System	58
26	Sunlight Radio Receiver	60
27	Powerful Two-Transistor Room Metronome	62
28	Battery Eliminator	64
29	Beginner's Regenerative Radio Set	66
30	Headset Booster	68
31	Transistorized Electronic Timer	70
32	Audio Filter Unit	72
33	Crazy Kiddie Toy	74
34	Sensitive Geiger Counter	76
35	Simplest Audio Amplifier	78
36	Sun-Powered Code Practice Oscillator	80
37	Transistorized Crystal Oscillator	82
38	Inexpensive Direct-Coupled Amplifier	84
39	House Wire CW Set	86
40	CW Montior	88
41	Hi-Fi Audio Mixer	90
42	Push-Pull Receiver	92
43	Personal Metronome	94
44	Two-Transistor Sensitive Light Relay	96
45	Blinker	98
46	Square-Wave Audio Generator	100
47	Automatic Sound Switch	102
48	Audio Frequency Meter	104
49	One-Transistor Light Relay	106
	Index	108

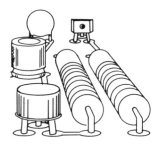

Preface

This book is a virtual cookbook of easy-to-build gadgets that you can assemble using only two types of transistors. Why only two transistors? Simply because not everyone has a pocketbook that permits his buying transistors and assorted components when he wants to start experimenting at the workbench.

First, you should build up a good supply of resistors and capacitors by buying a few standard assortments of these components. In fact, many can be salvaged from old radio and television chassis. (Be sure to check each component after removal.) If you do not have any nonworking radio and television sets, visit the repair shops in your neighborhood; many discarded chassis can be purchased for very little money. One word of caution, however; some parts, such as electrolytic capacitors, deteriorate with age. They may check good when removed from a discarded set but will not work when you try to use them.

All projects in this book are complete with descriptive texts, parts lists, and schematic diagrams. Only standard schematic symbols are used in the diagrams, and all capacitors and resistors are standard values. Tolerances are not critical in any of the projects but should be followed as closely as possible.

These projects have been designed to make maximum use of certain key parts. You will note that many parts, such as transistors, are used over and over again. Take advantage of this convenience by referring to the parts list in each project. You will discover just how fascinating and exciting the world of electronics can be—without spending much money on a profitable hobby.

Robert M. Brown
Tom Kneitel

Other TAB books by the authors:

No. 1339 *101 Easy Test Instrument Projects*
No. 1347 *49 More Easy-To-Build Electronics Projects*

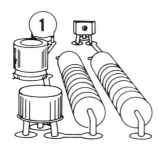

Wireless Home Broadcaster

A subminiature wireless microphone that can be built into an empty hard-pack cigarette box is our first project. Using conventional components, it provides both excellent modulation and a dandy little variable-frequency transmitter to plop your signals down just about anywhere in the standard 550-1650 kHz AM broadcast band.

The heart of the broadcaster is the oscillator circuit, which uses a 2N366 transistor in conjunction with a tuned circuit resonating in the AM band. Although we used an older antenna coil in this project and others, if you like you can use a tapped transmitter variable antenna coil in conjunction with a subminiature 365-pf variable tuning capacitor for L1. Coil L2 is simply 10 to 15 turns of hookup wire wound over the middle of L1.

A regular 1K to 2K-ohm impedance magnetic earphone is used as the microphone precluding the necessity of a matching transformer. Gain is adjusted by setting R4 to the desired level. If you are using the slug-tuned antenna coil, use the slug to "calibrate" C6 by adjusting the slug so that when C6 is fully meshed your signal is at the bottom of the band; when C6 is fully open, it should hit the top of the AM band. In practice, merely find a good unused frequency and zero in your home broadcaster by adjusting C6 until your signal is heard in the receiver. That is all you have to do and you are on the air.

The antenna can be a short length of stiff (No. 10 or No. 12) bare wire acting as a whip. Do not use a length of wire longer than approximately eight feet, or your signal will radiate further than is allowed presently under Part 15 of the Federal Communications Commission rules and regulations. Provided you comply with this distance limitation (maximum: approximately 300 feet), no license is required. See Fig. 1-1 and Table 1-1.

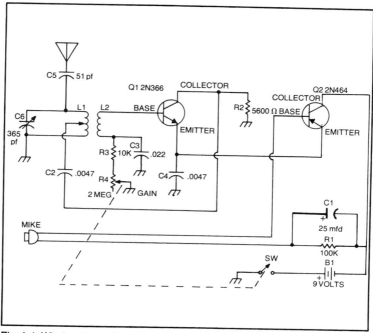

Fig. 1-1. Wireless home broadcaster circuit.

Table 1-1. Parts List for Wireless Home Broadcaster.

Item No.	Description
B1	9-volt battery.
C1	25-mfd, 15-volt electrolytic capacitor.
C2, C4	.0047-mfd capacitors.
C3	.022-mfd capacitor.
C5	51-pf capacitor.
C6	365-pf variable capacitor.
L1	Transistor antenna coil (Lafayette MS-166 or equiv.).
L2	10-15 turns of hookup wire around center of L1.
Q1	2N366 transistor.
Q2	2N464 tranisistor.
R1	100K resistor.
R2	5600-ohm resistor.
R3	10K resistor.
R4	2-meg potentiometer with spst switch.
	Earphone (1K to 2K magnetic type).

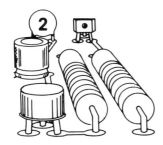

Hand Motion Music Maker

Here is a unique "mystery box" which when placed near an ordinary AM radio will produce wild musical sounds, and (with some skill) actually play melodies—all without you even touching the radio *or* the music maker.

The secret is in the body capacitance created by moving your hand near the music maker enclosure which, incidentally, should *not* be metal. As your hand gets closer and closer to the device, a strong musical note will suddenly burst from the radio, rising higher and higher in frequency as your hand gets closer to the music maker. By quavering your hand motion, an effective Hawaiian guitar sound is produced. When you remove your hand from the vicinity of the device, the music suddenly stops.

The heart of the unusual hand motion music maker is double transmitter circuit (two separate rf signals) which radiate to the nearby table or transistor radio by means of a large antenna loop coil. By carefully locating a blank spot on the lower portion of the radio dial, you adjust C2 until a signal is heard. Then, you tune the slug of L2 until you hear a gradual "zero-beating" taking place— meaning the two signals are nearing the same frequency on the dial. When you have made a perfect "hit," no sound will be heard at all, although you will know that a "station" is being tuned in. As you back away from the device, the tone will be heard.

At this point, use a little trial-and-error in adjusting both L2 and C2 to a point where no sound is heard (perfect zero-beat) when you are *not* near the gadget. Then, as your hand comes close to L2, it will alter the tuned circuitry just enough to cause a tone to burst from the radio speaker. See Fig. 2-1 and Table 2-1.

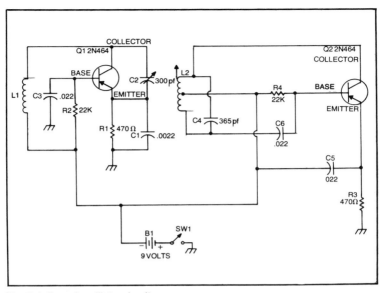

Fig. 2-1. Sound oscillator circuit.

Table 2-1. Parts List for Hand Motion Music Maker.

Item No.	Description
B1	9-volt battery.
C1	.0022-mfd capacitor.
C2	300-pf variable capacitor.
C3, C5, C6	.022-mfd capacitors.
C4	365-pf capacitor.
L1	Standard loop-type AM radio antenna coil. (This is normally found on the back of a-c/d-c radios.)
L2	Variable loopstick (Supervex VLT-240 or equiv.).
Q1, Q2	2N464 transistors.
R1, R3	470-ohm resistors.
R2, R4	22K resistors.
SW1	Spst switch.

13

Personal Code Practice Oscillator

Why pay a high price for a commercial code practice oscillator when with this circuit you can put one together with just a small handful of parts? Best of all, this oscillator matches perfectly to magnetic-type headphones, meaning that you can practice the International Morse Code for hours without disturbing others.

You can adjust the pitch of the CW notes to any desired frequency by adjusting the setting of R3. No power switch is required, since, unless the key is momentarily depressed, the circuit is not functioning and consequently will not draw current from the battery. If you like, you can feed this oscillator into a headset-to-speaker-adapter circuit (such as is described in Project 8) and wind up with room-filling sound. See Fig. 3-1 and Table 3-1.

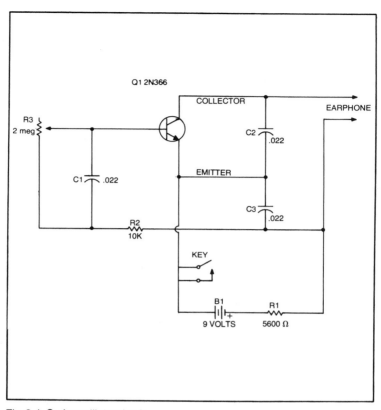

Fig. 3-1. Code oscillator circuit.

Table 3-1. Parts List for Code Practice Oscillator.

Item No.	Description
B1	9-volt battery.
C1, C2, C3	.022-mfd capacitors.
Q1	2N366 transistor.
R1	5600-ohm resistor.
R2	10K resistor.
R3	2-meg potentiometer.
	Key.
	Earphone (1K to 2K magnetic type).

One-Transistor AM Radio

Anyone who has ever had any experience with crystal radio sets knows all too well that although they are fun, they supply very little volume and are not very sensitive or selective. By combining a crystal detector, however, with an inexpensive transistor, you have a good little AM radio receiver that pumps an amazing amount of audio output to a pair of magnetic headphones. Best of all, it really pulls in distant stations—especially when it is provided with a good long-wire antenna strung between trees. But even with a 10-foot length of hookup wire serving as the antenna, it will do an admirable job and will run forever from one inexpensive 9-volt transistor battery.

The heart of the circuit is the L/C (resonant) section provided by the antenna coil and C1, a 365-pf variable capacitor. If you like, you can use a slug-tuned coil in place of L1 and forget altogether about needing a *variable* tuning capacitor. Instead if you use the newer coil, just drop in a tiny ceramic 60-pf capacitor and do all the tuning with L1. You can mount a small tuning knob on the screw-shaft of L1. The detector is a standard crystal diode, available for less than a dollar at most electronic parts stores. See Fig. 4-1 and Table 4-1.

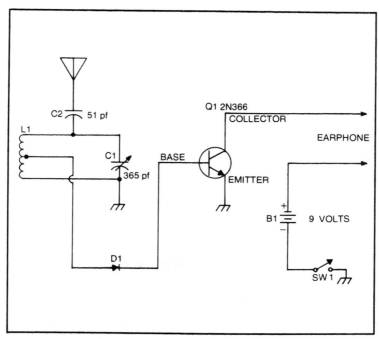

Fig. 4-1. One-transistor AM radio circuit.

Table 4-1. Parts List for One-Transistor AM Radio.

Item No.	Description
B1	9-volt battery.
C1	365-pf variable capacitor.
C2	51-pf capacitor.
D1	1N38B diode.
L1	Transistor antenna coil (Lafayette MS-166 or equiv.).
Q1	2N366 transistor.
SW1	Spst switch.
	Earphone (1K to 2K magnetic type).

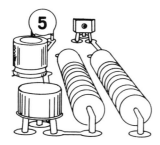

All-Purpose Signal Generator

Do you want a useful, but inexpensive, audio signal generator for checking out your audio projects? Or maybe you have a pile of nonworking radios and amplifiers that you would like to get working again? Regardless of your purposes, this handy little signal generator is just the ticket for all-around troubleshooting, and contains all the features found in the more expensive commercial versions.

It can be built into a small aluminum *Mini-box* and equipped with only two control knobs—R2 to adjust the pitch or audio frequency output, and R3 to serve as an off-on and "level" control. Once it is completed, you can calibrate your signal generator simply by comparing its output to that of a known source such as another generator. For about $1.00 or so, most radio and tv shops will calibrate your instrument for you in whatever frequency increments you desire. If you want to generate some off-beat frequencies, experiment a bit with the value of C1. See Fig. 5-1 and Table 5-1.

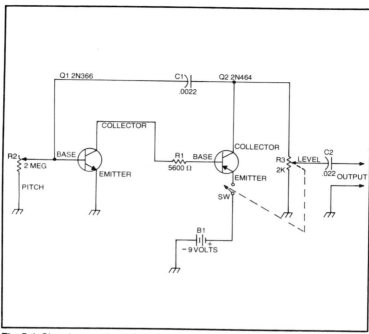

Fig. 5-1. Signal generator schematic.

Table 5-1. Parts List for Signal Generator.

Item No.	Description
B1	9-volt battery.
C1	.0022-mfd capacitor.
C2	.022-mfd capacitor.
Q1	2N366 transistor.
Q2	2N464 transistor.
R1	5600-ohm resistor.
R2	2-meg potentiometer.
R3	2K potentiometer with spst switch.
	Earphone.

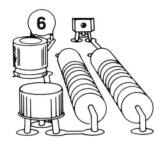

Echo-Chamber Amplifier

All you need to produce those exotic echo effects heard so often these days on psychedelic recordings and in television commercials is an inexpensive recording head, an inexpensive battery-powered tape recorder, and this easy-to-construct add-on circuit. The result will be a definite echo delay which you can insert in anything you wish to record.

The heart of the device is, of course, the added tape head. This must be placed on a mounting just to the right of the existing tape head in such a way as to ensure that the moving tape flows over *both* heads before feeding to the take-up reel. Do not worry about the tape head being expensive since you won't be concerned with recording—just playback. The output is fed to the "hot" lead on the volume control so that the resultant output will be mixed with that of the original recording.

Note that this circuit calls for a positive ground configuration. If after you examine your recorder schematic diagram you find that it utilizes a negative-ground arrangement, do not hook this adapter in; it must be changed accordingly. To accomplish this, reverse the battery and capacitive polarities (so that the negative terminals go to ground), and change the transistor to an NPN 2N366. See Fig. 6-1 and Table 6-1.

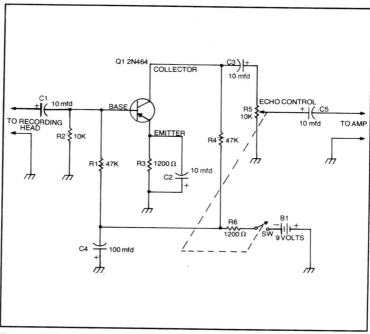

Fig. 6-1. Echo-chamber circuit.

Table 6-1. Parts List for Echo Chamber.

Item No.	Description
B1	9-volt battery.
C1, C2, C3, C5	10-mfd, 75-volt electrolytic capacitors.
C4	100-mfd, 35-volt electrolytic capacitor.
Q1	2N464 transistor.
R1, R4	47K resistors.
R2	10K resistor.
R3, R6	1200-ohm resistors.
R5	10K potentiometer with spst switch.
	Replacement-type inexpensive recording head.

21

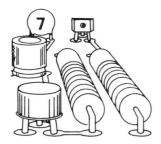

Automatic Safety Flasher

You have seen plenty of those so-called electronic flashers—but have you ever come across one which would *only come on at night?* If you have been doing outside construction work and do not want to worry about turning off your flasher lights, or about replacing dead batteries, take a peek at this simple circuit. It will flash practically forever on four size-D flashlight cells. It will furnish a bright nighttime illumination, and shut itself off automatically as soon as the sun comes up next morning.

For best effect, a bulb cover, purchased at a hardware store, will give your flasher a distinct orange, amber, or red glow—depending upon the type of cover your select. For a few cents more, you can mount a magnifying cover over your pilot bulb for a truly brilliant flashing effect.

The photocell must be mounted on top of the unit in such a way as to detect the greatest amount of available light, yet not accidentally get covered with dirt. It is always a good idea to check it from time to time to make sure that the surface is clean. If you use a large 6-volt battery, you will get a much more brilliant light. Or, if you like, you can use a 6-volt automobile bulb available at automotive supply stores. See Fig. 7-1 and Table 7-1.

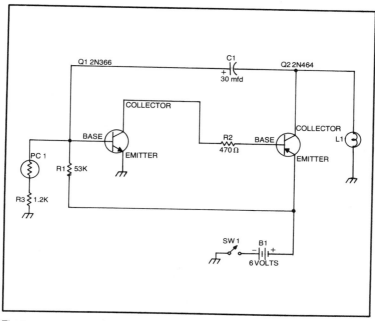

Fig. 7-1. Flasher circuit.

Table 7-1. Parts List for Automatic Safety Flasher.

Item No.	Description
B1	6-volt battery.
C1	30-mfd, 15-volt electrolytic capacitor.
I1	No. 47 type 6.3-volt ac pilot bulb.
PC1	Photocell (International Rectifier B2M or equiv.).
Q1	2N366 transistor.
Q2	2N464 transistor.
R1	53K resistor.
R2	470-ohm resistor.
R3	1.2K resistor.
SW1	Spst switch.

Headset-To-Speaker Adapter

Many of the projects in this book, including radio receivers, tone oscillators, and sound amplifiers, use conventional magnetic earphones for output purposes. With the addition, however, of this easy-to-construct one-transistor adapter, you can amplify these gadgets sufficiently to drive a small pm (Alnico-V) speaker, with amazing, roomfilling sound.

The adapter connects directly to the earphone terminals, requiring no matching transformers except at the speaker end. Requiring only a handful of parts and an inexpensive transistor, the adapter can be built into a small plastic parts box or aluminum *Mini-box*. If you like, you can built it right into your pm speaker box.

The only protruding control knob is R1-SW, which acts as an on-off/volume control. Your adapter amplifier is powered by two 6-volt batteries, mounted in an appropriate series of battery clips. Any conventional 3.2, 4, or 8-ohm pm speaker will do nicely. See Fig. 8-1 and Table 8-1.

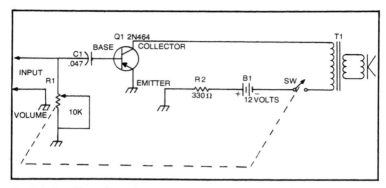

Fig. 8-1. Amplifier schematic.

Table 8-1. Parts List for Headset-to-Speaker Adapter.

Item No.	Description
B1	12-volt power source (two 6-volt batteries).
C1	.047-mfd capacitor.
Q1	2N464 transistor.
R1	10K potentiometer with spst switch.
R2	330-ohm resistor.
T1	Transistor output transformer (Argonne AR-167 or equiv.).
	Speaker.

Batteryless Transistor Receiver

Here is perhaps the ultimate in transistor-circuit simplicity: a sensitive AM radio receiver requiring only four components plus an earphone and antenna wire. Using a transistor, loopstick antenna coil, a coil link wound over the middle of L1, and a subminiature transistor-type 365-pf variable tuning capacitor, the little receiver can literally be built into a small pillbox.

The trick to success is to employ a 10-foot length of insulated hookup wire equipped with an appropriate-size alligator clip as the antenna. In this manner, you can clip your little receiver to any large metallic object that will serve as a giant antenna. The best bet is to connect your alligator clip to the metal finger stop on the household dial telephone. The resultant increase in signal intensity will often be sufficient to drive a speaker.

Calibrate your tiny receiver by adjusting the screw-down ferrite slug on L1 so that the bottom of the AM band is received (approximately 550-600 kHz) when the capacitive plates of C1 are fully meshed. Conversely, you should receive the top of the band signals when the plates of C1 are fully unmeshed (approximately 1650-1700 kHz). Once L1 has been calibrated, you can leave it and do all future tuning of the band with capacitor C1. Incidentally, discarded transistor radios are great for yielding C1 capacitors. See Fig. 9-1 and Table 9-1.

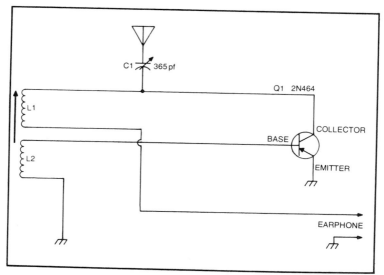

Fig. 9-1. AM radio circuit.

Table 9-1. Parts List for Batteryless Transistor Receiver.

Item No.	Description
C1	365-pf variable capacitor.
L1	Variable loopstick antenna coil (Superex VLT-240 or equiv.)
L2	6 turns of No. 22 enameled wire wound over the middle of L1.
Q1	2N464 transistor.
	Earphone.

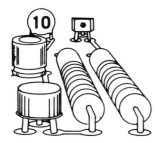

Magic Meter Sounder

One of the nearest things to perpetual motion is this unusual gadget. It not only produces a tone, but it also causes a meter to deflect at the same time. So what is different about that? Simply that after you turn the sounder on, it will continue to turn *itself* on and off at the rate of approximately once each second. This means simply that you will have a perpetual beeper and meter deflector—an admittedly useless do-nothing box—but one which makes a dandy science fair project.

The purpose of R1-SW is to permit you to apply initial power to the unit in addition to adjusting the triggering of the automatic on-and-off action. Vary the setting of R1 until the earphones indicate that the sounder has started to oscillate. Now wait, and in a second or so, it will stop. Wait another second, and the meter will rise and a second tone will be heard in the earphone. The automatic meter sounder will continue to operate like this for months before requiring a new battery. See Fig. 10-1 and Table 10-1.

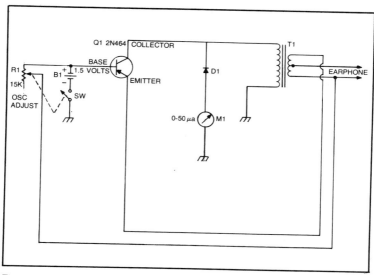

Fig. 10-1. Beeper circuit.

Table 10-1. Parts List for Meter Sounder.

Item No.	Description
B1	1.5-volt dry cell.
D1	1N38B diode.
M1	0-50 d-c microammeter.
Q1	2N464 transistor.
R1	15K potentiometer with spst switch.
T1	Transistor output transformer (Argonne AR-103 or equiv.).
	Earphone (1K to 2K magnetic type).

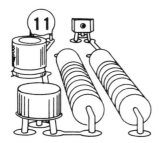

Amazing Electronic Music Maker

Known affectionately as an audio light detector, this unusual gadget converts available light into music. All you do is place it near a lamp, and the speaker emits tones instantly. To play songs, simply wave your hand between the light source and the photocell. By interrupting the light beams in this manner, you are varying the intensity of light reaching the sensitive photocell which in turn develops different audio tones. A little practice, and you will soon be playing familiar melodies with ease.

Technically speaking, this is *not* a light-powered oscillator. It uses its own self-contained 9-volt battery to power the speaker and transistors. Only the bias resistance of Q1 is altered by the photocell. The result is an instrument that produces a wide variety of tones, but at the same volume level. Incidentally, you will find that the greater the light source (for example, direct sunlight), the higher the resultant pitch. Allowing the light to pass through your fingers is an excellent way of diffusing it. See Fig. 11-1 and Table 11-1.

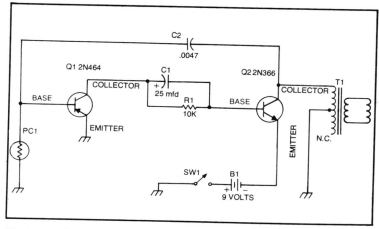

Fig. 11-1. Electronic music maker circuit.

Table 11-1. Parts List for Electronic Music Maker.

Item No.	Description
B1	9-volt battery.
C1	25-mfd, 15-volt electrolytic capacitor.
C2	.0047- mfd capacitor.
PC1	Photocell (International Rectifier B2M or equiv.).
Q1	2N464 transistor.
Q2	2N366 transistor.
R1	10K resistor.
SW1	Spst switch.
T1	Audio output transformer (Argonne AR-170 or equiv.).
	Speaker.

Miniature FM Radio

Here is an interesting little receiver which—believe it or not—will tune the entire 88-108 MHz FM music band, yet will provide sufficient output to drive a standard pair of magnetic headphones. So small it can be built into an empty cigarette pack, the tiny receiver can also pickup the sound tracks of certain tv stations.

The only tricky part is the L/C circuit, which consists of a 9-pf trimmer capacitor and a tapped antenna coil in parallel. First, try out the circuit after constructing L1. If you do not hear the FM stations, try reversing the taps on the coil. If you still have difficulty, experiment a bit with a coil by moving the taps slightly, one-half-turn at a time, separately—listening carefully on the earphones while you adjust C1 with each change you make.

The antenna can be almost any length of stiff (whip-like) No. 10 or No. 12 wire, although we found that 3⅛ inches was perfect for the middle of the FM band. A vhf diode is used in case you cannot obtain a 1N82. Types such as the 1N21 are also good at this frequency. If magnetic earphones (which we recommend) are used, eliminate R1 entirely from the circuit. You may be able to connect to a crystal (transistor radio type) earphone in the circuit if R1 is used. See Fig. 12-1 and Table 12-1.

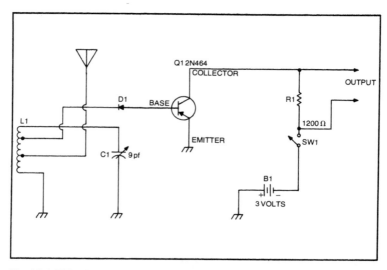

Fig. 12-1. FM radio circuit.

Table 12-1. Parts List for FM Radio.

Item No.	Description
B1	3-volt power source (two 1.5-volt dry cells).
C1	9-pf variable trimmer capacitor.
D1	1N82 diode.
L1	5 turns of No. 16 enameled wire wound on a ¾-inch diameter coil form, ½ inch long. Tapped at ½ turn from ground end for antenna and 2 turns from top of coil for diode.
Q1	2N464 transistor.
R1	1200-ohm resistor.
SW1	Spst switch.
	Earphone (magnetic).

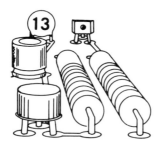

CB Field Strength Meter

For the uninitiated, an fsm is a *field strength meter*—a gadget designed to measure the actual rf output of a given transmitter. With the design shown here, you can accurately determine the point of maximum transmitter output of any CB transceiver, including nonlicensed walkie-talkies.

Merely bring the 1-inch antenna wire (whip) close to the CB antenna and adjust R1 for a one-third- to one-half-scale meter reading on M1. If the needle appears to be going off scale to the left, reverse the meter connections. Adjust the antenna tuning capacitor on the CB set for maximum reading on M1. When a point is reached where you can no longer cause the meter to rise, you have achieved optimum output and best antenna impedance matching.

For greatest sensitivity for weak-signal sources such as low-power walkie-talkies, tune C1 to the exact operating frequency. This will be signaled by a definite and sudden rise on M1 when you hit this frequency. Once this has been preset, you can proceed with your tuneup as described previously. The coil is made from a prefabricated *Mini-ductor* form by cutting off a 12-turn section to use as L1. See Fig 13-1 and Table 13-1.

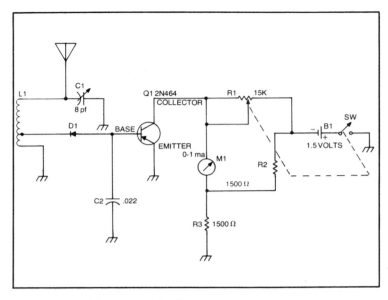

Fig. 13-1. Field strength meter circuit.

Table 13-1. Parts List for Field Strength Meter.

Item No.	Description
B1	1.5-volt dry cell.
C1	8-pf variable trimmer capacitor.
C2	.022-mfd capacitor.
D1	1N34A diode.
L1	12 turns of B&W No. 3015, tapped at 3 turns from the ground end.
M1	0-1 dc milliammeter.
Q1	2N464 transistor.
R1	15K potentiometer with spst switch.
R2, R3	1500-ohm resistors.

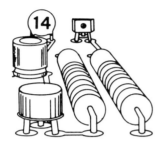

Dynamic Mike

Are you aware of the fact that you can use a small pm speaker as a powerful microphone? It can be done and is in fact the method the walkie-talkie manufacturers employ to allow you to talk directly into the speaker in reply to a call.

Actually, you can use any size speaker to build your own dynamic mike. Generally speaking, the larger the speaker, the greater the fidelity of the resultant output. Conversely, the smaller the speaker, the greater the loss in fidelity. In most applications, however, a small pm speaker from a discarded transistor radio makes a great microphone.

Note carefully the polarity arrangements shown in Fig. 14-1. The negative terminal of the battery (B2) is switched to ground through SW1B while the positive electrode of B1 goes to ground through SW1A. No matching transformer is necessary to couple the speaker to the NPN transistor. Also, see Table 14-1.

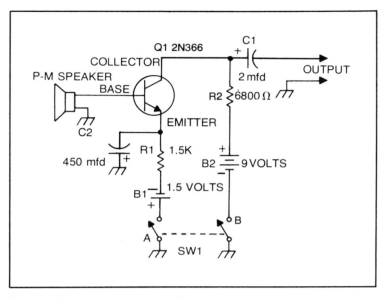

Fig. 14-1. Dynamic microphone circuit.

Table 14-1. Parts List for Dynamic Mike.

Item No.	Description
B1	1.5-volt dry cell.
B2	9-volt battery.
C1	2-mfd, 15-volt electrolytic capacitor.
C2	4 50-mfd, 15-volt electrolytic capacitor.
Q1	2N366 transistor.
R1	1.5K resistor.
R2	6800-ohm resistor.
SW1	Dpst switch.
	Speaker.

Handy Signal Tracer

Do you want to do a little troubleshooting on broken down electronic devices? If you do, you will need a good signal tracer, a device capable of amplifying minute amounts of audio to a point where they can be "read out" on a pair of monitoring headphones. Such a gadget is shown in Fig. 15-1.

Let's say, for example, that you *think* the trouble with a record player amplifier is in the audio output circuit, but you do not know just where. By placing the tracer on the amplifier and probing with the "hot" tracer input lead you will be able to pinpoint the difficulty. Suppose, for example, that you hear the recorded music up to and including a lead that feeds the final push-pull output stage, but you somehow lose the audio after this point. You spot a .01-mfd bypass capacitor running from that lead to ground. You check the audio after this capacitor and get nothing so you disconnect the grounded lead of the .01-mfd and monitor with the tracer probe connected to the same point. This time you have tremendous audio—so strong in fact that you have to readjust R2 before you damage your eardrums.

In this manner you have found the trouble: a defective bypass capacitor that was shorting nearly all the audio signal to ground. With the tracer you were able to pinpoint the problem, something that would have taken needless hours without such an aid. Remember always to start your troubleshooting at the input section of whatever audio equipment is defective. Slowly trace through the entire audio-carrying circuit until you find a point where there is no amplification. Also, see Table 15-1.

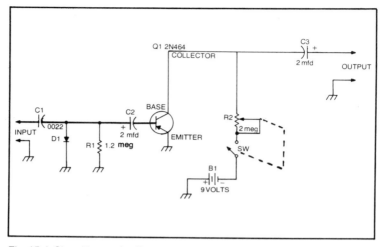

Fig. 15-1. Signal tracer circuit.

Table 15-1. Parts List for Signal Tracer.

Item No.	Description
B1	9-volt battery.
C1	.0022-mfd capacitor.
C2, C3	2-mfd, 15-volt electrolytic capacitors.
D1	1N38B diode.
Q1	2N464 transistor.
R1	1.2-meg resistor.
R2	2-meg potentiometer with spst switch.
	Speaker.

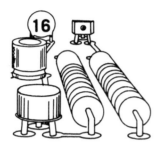

Legal CW (Code) Radio Transmitter

Would you like to go on the air an hour from now sending coded messages to your buddy across the street? You can accomplish this easily with the circuit shown in the accompanying diagram. All you need to receive the signal is a standard transistor radio receiver tuned to a blank space on the band. In operation, you simply adjust C1 until you hear a high-pitched tone in the monitoring receiver when the key is depressed. If you are not interested in fancy sending, you can substitute an inexpensive normally-open pushbutton switch for the telegraph key.

Adjust the setting of R1 with the key depressed until the desired pitch is obtained. Note that this is *not* a beat-frequency gadget, but a complete modulated transmitter. Oscillation takes place by means of regeneration, and tone generation is caused by the blocking action of Q1. Keep your transmit antenna fairly short; otherwise you will be heard blocks away and the FCC will soon be after you. For more information on the limitations, check the last paragraph in Project 1. Incidentally a standard loopstick can be used instead of L1. Just tune with the variable slug, and replace C1 with a fixed 360-pf ceramic capacitor. See Fig. 16-1 and Table 16-1.

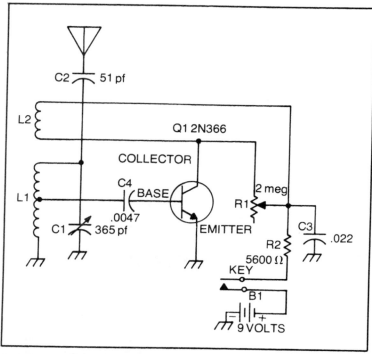

Fig. 16-1. Code transmitter circuit.

Table 16-1. Parts List for Radio Transmitter.

Item No.	Description
B1	9-volt battery.
C1	365-pf variable capacitor.
C2	51-pf capacitor.
C3	.022-mfd capacitor.
C4	.0047-mfd capacitor.
L1	Transistor antenna coil (Lafayette MS-166 or equiv.).
L2	10-15 turns of hookup wire wound over the middle of L1.
Q1	**2N366 transistor**
R1	2-meg potentiometer.
R2	5600-ohm resistor.
	Key.

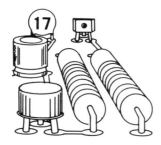

High-Power AM Receiver

While the majority of transistor radio receiver projects are a lot of fun to build, the greatest number of these lack something when it comes to performance. If you build the receiver in the accompanying diagram, you will be in for a shock. It has a supersensitive antenna circuit that will bring in most distant AM stations, plus a powerful audio amplifier that will even power a speaker when the receiver is tuned to local stations. In general use, however, it has been designed to pump plenty of music and news into a standard pair of 1K to 2K-ohm magnetic headsets.

Using a regenerative input circuit, signals will be picked up so fast that you may even have trouble separating them. Generally, however, you will be doing most of your tuning with C1. If you would rather use a loopstick in place of L1, you can replace C1 with a fixed 360-pf ceramic capacitor and tune with a small knob fixed to the shaft of the tuning slug of the loopstick antenna coil. In any case, adjust R1 for best regenerative kick-in (bringing with it the best of 2-transistor AM reception), and leave it there. R1 is *not* a volume control. See Fig. 17-1 and Table 17-1.

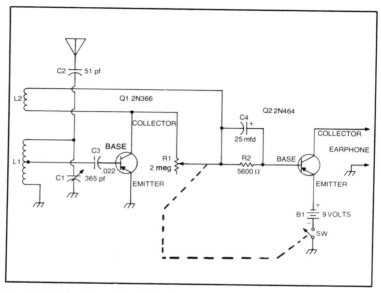

Fig. 17-1. AM receiver schematic.

Table 17-1. Parts List for Radio Receiver.

Item No.	Description
B1	9-volt battery.
C1	365-pf variable capacitor.
C2	51-pf capacitor.
C3	.022-mfd capacitor.
C4	25-mfd, 15-volt electrolytic capacitor.
L1	Transistor antenna coil (Lafayette MS-166 or equiv.).
L2	10-15 turns of hookup wire wound over the middle of L1.
Q1	2N366 transistor.
Q2	2N464 transistor.
R1	2-meg potentiometer with spst switch.
R2	5600-ohm resistor.
	Earphone.

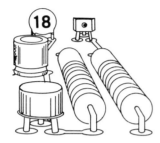

80 To 150 MHz VHF Receiver

Do you ever have an uncontrollable urge to know what is going on in the little-known vhf areas? You are in for a treat because this pocket-sized receiver will pick up such broadcasts as FM music, airplane pilots and control towers, military affilliate radio system, and 2-meter hams. Best of all, the whole thing can be built for less than $5.00.

The heart of this vhf eavesdropper is the antenna tuned circuit. If you construct the coil exactly as described, little trouble should be encountered. If you like, however, you can substitute your own design, providing you are handy with a grid dip meter to ensure that the same general 80-150-MHz resonancy can be obtained. We used an 8-pf trimmer capacitor in conjunction with the coil specifications obtained, although a standard miniature variable of the same rating will prove more versatile. The trimmer requires screwdriver tuning. The output can be coupled directly to a standard pair of magnetic earphones. Do not substitute a 1N34A diode for the 1N82A vhf type. A 1N21 surplus type can be used. See Fig. 18-1 and Table 18-1.

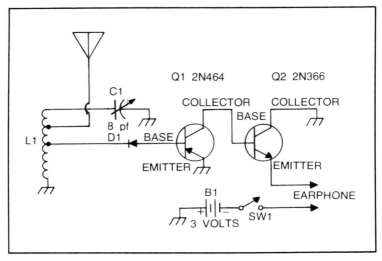

Fig. 18-1. VHF receiver circuit.

Table 18-1. Parts List for a VHF Receiver.

Item No.	Description
B1	3-volt battery.
C1	8-pf variable trimmer capacitor.
D1	1N82 diode.
L1	4 turns of No. 16 tinned wire on a ¾-inch diameter coil form, spaced ½ inch long. Antenna tap is located ½ turn from the high end. Diode tap is located 2 turns up from the ground end.
Q1	2N464 transistor.
Q2	2N366 transistor.
SW1	Spst switch.
	Earphone.

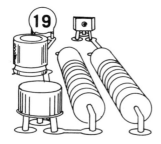

Supersonic Eavesdropper

An invisible-wires transmission system? You bet, and you don't even need a radio transmitter to do it! This is possible through an interesting phenomenon known as supersonic transmission—roughly a means of no-wires communications similar to induction methods.

The heart of the supersonic receiver is L1, a conventional telephone pickup coil. The coil receives the signal and sends it to Q1 and Q2 which in turn amplify it and feed it to a pair of monitoring headphones. Your transmitter can be any kind of audio amplifier rigged in the following fashion: run a gigantic "loop antenna" around the room, returning the end of the wire to the amplifier. Disconnect the speaker wires and connect the two leads of the continuous "loop" across this point. Now, measure the impedance of your supersonic transmission line with an ohmmeter. If you are using an 8-ohm speaker system and the loop only registers a total impedance of 2 ohms, insert a 6-ohm resistor (of wattage compatible with that of your amplifier output) in series with the loop. Once this impedance-matching procedure has been accomplished, you are ready for action.

By simply pocketing the supersonic eavesdropper, you will now be able to clearly hear everything being played over the audio amplifier system—although nobody else will be in on the secret. If the resultant volume is too loud, merely adjust R4 accordingly.

Incidentally, a system similar to this is being used now in the New York City Museum of Natural History. Vistors insert monitoring earpieces into their ears and carry supersonic eavesdropping receivers in their pockets. While they view the exhibits, a recorded voice explains what is going on. The more powerful your amplifier, the greater the overall range capability. See Fig. 19-1 and Table 19-1.

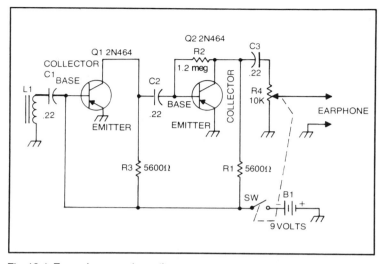

Fig. 19-1. Eavesdropper schematic.

Table 19-1. Parts List for Supersonic Eavesdropper.

Item No.	Description
B1	9-volt battery.
C1, C2, C3	.22-mfd capacitors.
L1	Standard telephone pickup coil.
Q1, Q2	2N464 transistors.
R1, R3	5600-ohm resistors.
R2	1.2-meg resistor.
R4	10K potentiometer with spst switch.
	Earphone.

Powerful Code Practice Monitor

If you are studying Morse code with a friend, those "personal" cpo's are *out*. You need an oscillator that will provide adequate volume for both of you to hear, yet still be as inexpensive as possible to put together. Such a gadget is shown in the accompanying diagram.

Using a handful of resistors, a battery, key, one capacitor, an inexpensive output transformer and transistor, this circuit provides enough output to drive any conventional pm speaker. Generally speaking, the smaller the speaker you use, the louder will be the sound.

The entire affair can be built right into the speaker enclosure itself, although we obtained all the parts including the speaker, earphone jacks, and off-on-volume control from a discarded transistor radio. This left us with plenty of space for the battery and associated components. The "key" leads were fed to the earphone jack. When we want to practice CW, all we have to do is plug in the key. No off-on switch is required since the circuit draws no power until the key is depressed. See Fig. 20-1 and Table 20-1.

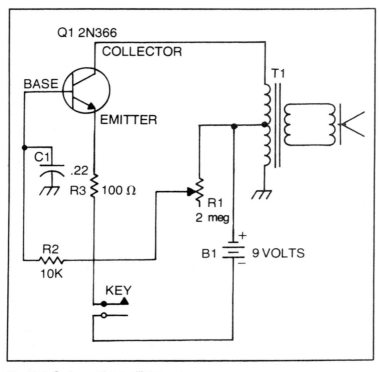

Fig. 20-1. Code practice oscillator.

Table 20-1. Parts List for Code Practice Monitor.

Item No.	Description
B1	9-volt battery.
C1	.22-mfd capacitor.
Q1	2N366 transistor.
R1	2-meg potentiometer.
R2	10K resistor.
R3	100-ohm resistor.
T1	Audio output transformer (Argonne AR-119 or equiv.).
	Key.
	Speaker.

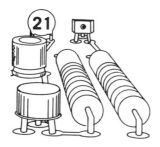

Automatic AC Drive Controller

Have you ever wanted to automatically turn off high-power ac appliances remotely from a variety of sources—such as manual switches, temperature-sensitive thermistors, moisture sensors, and even low-current photocells? You can do it with ease using the unique circuit shown in the accompanying diagram. All you do is plug your ac device (electric heater, dehumidifier, washing machine, etc.) into the 117-volt ac socket and connect your sensor to the input terminals. A direct short—such as would be provided if you hooked a switch across the input—is not necessary; the controller will trigger with high resistance values.

In operation, adjust R3 to "fire" the appliance on at the desired sensitivity point of the sensor. What happens is: as the proper resistance point is reached, low-current relay K1 will close, causing the larger 117-volt ac relay (K2) to do the same. This will automatically shut off any appliance rated up to 775 watts you have plugged into the device. Care should be exercised during wiring to prevent shorts and also to prevent the ac line from accidentally coming in contact with the chassis. Bulb I1 will signal when the appliance has been turned off. See Fig. 21-1 and Table 21-1.

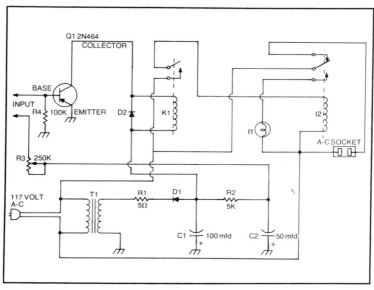

Fig. 21-1. Automatic AC control circuit.

Table 21-1. Parts List for AC Drive Controller.

Item No.	Description
C1	100-mfd, 15-volt electrolytic capacitor.
C2	50-mfd, 15-volt electrolytic capacitor.
D1	1N536 silicon.
D2	1N2069 silicon.
I1	5-watt, standard 117-volt ac household bulb.
K1	Spst 550-ohm relay (Sigma 11F-550G/SIL or equiv.).
K2	Spdt 117-volt ac relay (Potter & Brumfield MR-5A or equiv.).
Q1	2N464 transistor.
R1	5-ohm, 2-watt resistor.
R2	5K, 2-watt resistor.
R3	250K potentiometer.
R4	100K resistor.
T1	6.3-volt ac filament transformer (Triad F-14X or equiv.).

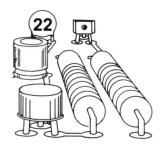

Efficient 2-Transistor Audio Preamplifier

Are you looking for a really powerful audio preamplifier that doesn't present a big construction job, yet can be put together with little time and expense? The circuit in the accompanying diagram represents an extremely high-gain audio preamplifier that can be hooked between a phono cartridge and a main amplifier for increased performance and high overall output.

The preamp should be built into an aluminum *Mini-box* (completely housed in this fashion) and equipped with appropriate phono input and output jacks plus an spst slide switch. Inside the box shielded cabling should be used on all leads carrying audio frequencies, particularly in the vicinity of input transistor Q1, the first 2N464. All leads should be kept very short and to the point. Watch battery polarity and the positioning of all electrolytic capacitors.

With proper care, no ac hum should be introduced. Incidentally, this circuit is also very effective as a microphone preamp. See Fig. 22-1 and Table 22-1.

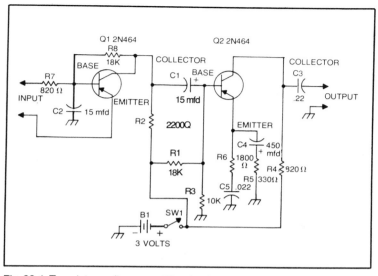

Fig. 22-1. Transistor audio preamplifier circuit.

Table 22-1. Parts List for Audio Preamplifier.

Item No.	Description
B1	3-volt battery.
C1, C2	15-mfd, 15-volt electrolytic capacitors.
C3	.22-mfd capacitor.
C4	450-mfd, 15-volt electrolytic capacitor.
C5	.022-mfd capacitor.
Q1, Q2	2N464 transistors.
R1, R8	18K resistors.
R2	2200-ohm resistor.
R3	10K resistor.
R4, R7	820-ohm resistors.
R5	330-ohm resistor.
R6	1800-ohm resistor.
SW1	Spst switch.
	Speaker.

Automatic Auto Light Reminder

Have you ever left your headlights on and run your battery down? This gadget is designed to let you know (by an audio blast) if you accidently leave your lights on after you turn off the ignition. Turn the lights off, and the reminder alarm blast will stop. That's all there is to it.

The circuit has been designed to work from virtually any vehicle—6 to 12 volts. It is entirely self-contained in a small aluminum box and can be mounted inconspicuously under the dashboard or even in the glove compartment. Hookup is easy. Connect the "ignition" lead to a point on the ignition switch that is also connected to the ignition coil. Next, connect the "lights" wire to the light switch terminal, the point that is connected to the taillights.

Now, make certain that the chassis of your reminder box is grounded to the body of the car—and you're in business. Incidentally, if you would like a cutout switch, you can install a simple spst slide switch in the "light" switch input lead. See Fig. 23-1 and Table 23-1.

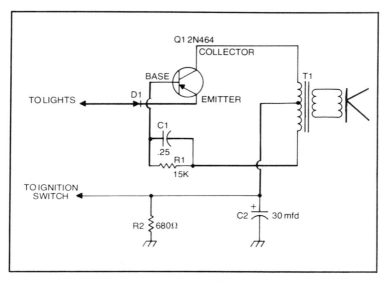

Fig. 23-1. Automatic headlight reminder circuit.

Table 23-1. Parts List for Auto Light Reminder.

Item No.	Description
C1	.25-mfd capacitor.
C2	30-mfd, 25-volt electrolytic capacitor.
D1	1N38B diode.
Q1	2N464 transistor.
R1	15K resistor.
R2	680-ohm resistor.
T1	Output transformer, 400-ohm centertap primary; 11-ohm secondary. Speaker.

Electronic Moisture/Rain Alarm

Would you like to know when to turn off the sprinkling system? Or maybe you would just like to be alerted when the basement has reached such a dampness point that it is time for the dehumidifier to come on? Better yet, how about switching the dehumidifier on automatically when it gets too damp? The applications of this gadget are limited only by your imagination.

The heart of the moisture/rain alarm is the moisture sensing element, MS1. This can be one of several different types currently sold by many parts stores. You can buy one which will detect fine changes in moisture content in the atmosphere, or others which require actual raindrops to trigger the action. In any case, hooking a moisture sensor into the circuit as shown will provide a solid switching function through relay K1 which can be used to turn devices on or off, depending on how you hook to the relay. Sensitivity—or "triggering point"—can be preset by simply adjusting potentiometer R4 to the desired level. See Fig. 24-1 and Table 24-1.

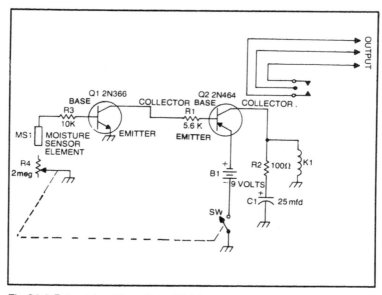

Fig. 24-1. Rain and moisture alarm circuit.

Table 24-1. Parts List for Moisture/Rain Alarm.

Item No.	Description
B1	9-volt battery.
C1	25-mfd, 15-volt electrolytic capacitor.
K1	4000-ohm spdt relay (Advance SO/IC/4000D or equiv.).
MS1	Moisture sensor element.
Q1	2N366 transistor.
Q2	2N464 transistor.
R1	5.6K resistor.
R2	100-ohm resistor.
R3	10K resistor.
R4	2-meg potentiometer with spst switch.

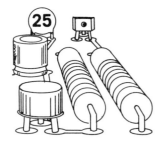

Light Beam Communications System

Do you want to send secret messages without wires over substantial distances without a radio transmitter or receiver? It is easy with this gadget—a sensitive light beam receiver that will "beep" whenever the proper "Signal" has been received.

Here is how it works: The photocell (PC1) is mounted vertically in your window and a cardboard tube is fitted over it—resulting in something that looks a great deal like a directional long-distance microphone. By line-of-sight trial-and-error, your buddy carefully positions a high-powered (three-cell or more) flashlight in his window so that the beam hits the cardboard tube exactly in the center. Connections that are made to bypass the on-off switch on the flashlight are run down to a telegraph key at his desk. Every time he hits the key, you will hear a "beep" in the headset. For a true-two-way system, identical sending and receiving systems are made for both operators.

You can use International Morse code or make up your own coding system. To provide added sensitivity, R2 can be adjusted so the photocell will "trigger" at the slightest amount of received light. This adjustment will also permit you to adjust the resultant audio tone as stronger light beams are received. Always be sure to turn SW1 off when the receiver is not being used. See Fig. 25-1 and Table 25-1.

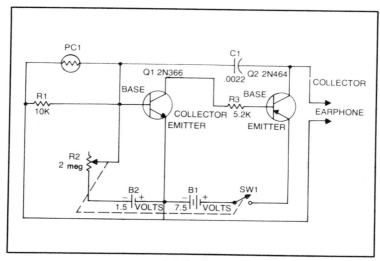

Fig. 25-1. Light beam circuit.

Table 25-1. Parts List for Communication System.

Item No.	Description
B1	7.5-volt battery.
B2	1.5-volt dry cell.
C1	.0022-mfd capacitor.
PC1	Photocell (International Rectifier B2M or equiv.).
Q1	2N366 transistor.
Q2	2N464 transistor.
R1	10K resistor.
R2	2-meg potentiometer with spst switch.
R3	5.2K resistor.
	Earphone.

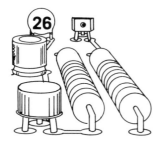

Sunlight Radio Receiver

Here is an unusual transistor radio receiver that is powered entirely by the sun. It is so small that it can be constructed in a tiny pillbox, but the receiver is much more sensitive than a crystal radio. When intense light hits the photocell, more volume is produced. When available incandescent light hits the sensitive cell, only local stations will be heard.

As with earlier projects, a variable loopstick antenna coil can be substituted for L1, although a fixed value 360-pf capacitor should be substituted for C1 if this is done. With such a replacement, all tuning would then be accomplished by tuning the slug of the loopstick.

With the transistor antenna coil, tuning is accomplished by varying the capacity of C1, a 365-pf subminiature variable. The antenna itself can be any long wire, or just a piece of insulated hookup wire with an alligator clip in the end. The alligator clip can then be affixed to any large metallic object (window frame, bedspring, etc.) which can then serve as a large antenna.

Incidentally, the set will work quite well when placed directly under a 75-watt table lamp. For truly great reception, move a Tensor lamp into position directly over the photocell and watch what happens. See Fig. 26-1 and Table 26-1.

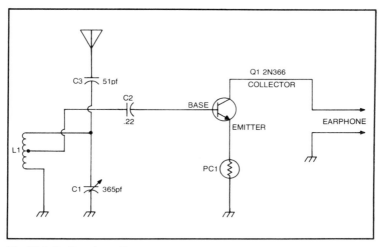

Fig. 26-1. Radio schematic.

Table 26-1. Parts List for Sunlight Radio Receiver.

Item No.	Description
C1	365-pf variable capacitor.
C2	.22-mfd capacitor.
C3	51-pf capacitor.
L1	Transistor antenna coil (Lafayette MS-166 or equiv.).
PC1	Photocell (International Rectifier B2M or equiv.).
Q1	2N366 transistor.
	Earphone.

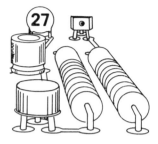

Powerful Two-Transistor Room Metronome

Clack-clack-clack-clack—that's what you'll hear in perfect synchronization with your piano (assuming the player has a good ear) when you turn this gadget on. The circuit is similar to that used in expensive electronic metronomes. They may look a lot nicer, but the circuit is almost identical.

Whenever you turn SW1 on, the clacking will begin, coming loudly from the speaker—yet not so loudly as to override the music. To adjust for perfect beat-setting, you merely set potentiometer R3 for the proper rhythm. Best of all, the entire affair can be built into a handsome wooden speaker box and is completely portable. Only one control protrudes from the front or side panel. See Fig. 27-1 and Table 27-1.

Incidentally, this gadget makes a great rain simulator for those who have difficulty getting to sleep at night. The constant clacking has an interestingly hypnotic effect on listeners.

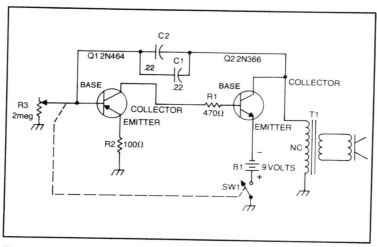

Fig. 27-1. Metronome circuit.

Table 27-1. Parts List for Room Metronome.

Item No.	Description
B1	9-volt battery.
C1, C2	.22-mfd capacitors.
Q1	2N464 transistor.
Q2	2N366 transistor.
R1	470-ohm resistor.
R2	100-ohm resistor.
R3	2-meg potentiometer with spst switch.
T1	Output transformer (Argonne AR-119 or equiv.).
	Speaker.

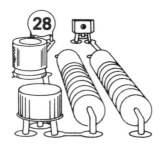

Battery Eliminator

By now you have probably noticed that a great deal of the projects in this book require a 6-volt dc supply for a power source. Others than call for a 9-volt dc supply will work in most cases just as well with 6 volts. In any case, here is an inexpensive transistorized battery eliminator that converts an ordinary 6.3-volt ac filament transformer into a good dc power supply that can also be lowered to 3 volts.

Unlike most power supply configurations, all you need for this one is a handful of parts, an inexpensive transistor, and a crystal-radio diode. The diode acts to rectify the ac from the transformer.

The output can be regulated by control R2, which serves to adjust the output voltage. You can adjust the voltage all the way down to 0 if you like, although it has been designed for 6-volt primary output.

The unit should be built in a small plastic box or aluminum *Mini-box*. Care should be exercised to prevent any shorting—particularly near the power transformer. Note output polarities during hookup to your projects. See Fig. 28-1 and Table 28-1.

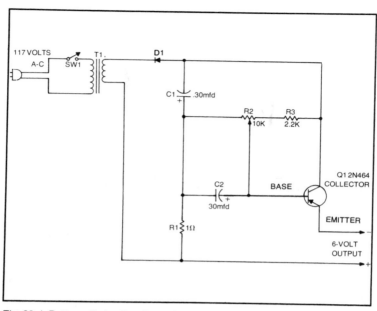

Fig. 28-1. Battery eliminator schematic.

Table 28-1. Parts List for Battery Eliminator.

Item No.	Description
C1, C2	30-mfd, 15-volt electrolytic capacitors.
D1	1N34A diode.
Q1	2N464 transistor.
R1	1-ohm resistor.
R2	10K potentiometer.
R3	2.2K resistor.
SW1	Spst switch.
T1	Standard 6.3-volt filament transformer.

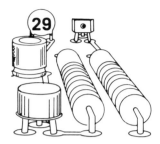

Beginner's Regenerative Radio Set

Anyone who has not yet sampled the "miracle" of regenerative reception has a pleasant surprise in store—a unique method of receiving signals that can make a one-transistor set perform like a three- or four-transistor commercial receiver. As you are probably aware, the problem with most simple transistor receivers is their lack of apparent sensitivity and volume, unless hooked up to a long wire antenna (preferably about 3 miles in length!) and solid earth ground.

A glance at the circuit in Fig. 29-1 however, shows what can be done with a minimum of components using the regenerative principle. Note particularly the added loop over the main antenna coil (L2 in the diagram). Potentiometer R1 controls the regeneration and remains pretty much as a preset adjustment. In operation, you adjust this control until a definite "rushing" sound is heard— the exact point of regenerative oscillation. This rushing sound is the clue that you are also bringing in hard-to-receive signals; to test, just tune capacitor C1 to a known station. Don't get carried away with R1; it is *not* a volume control and should not be advanced further than is necessary to bring in the rushing. If you are not careful with this adjustment (and also the L2 loop), you will set the whole affair into actual oscillation and wind up with something that sounds like a code practice oscillator.

As with some earlier projects, if you replace C1 with a 360-pf fixed ceramic, you can substitute a *variable* loopstick for L1 and do all of your tuning with the slug on L1. This is optional, however. Since we are dealing with a regenerative principle, it might be wise to tune with C1 since hand capacitance may upset the inductance ratios and set the receiver into oscillation. In any case, a bit of trial-and-error may be in order. Once the proper L1-L2 relation-

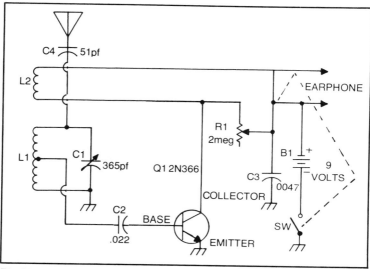

Fig. 29-1. Radio circuit.

ship has been established, a small amount of cement will hold things in place nicely.

The antenna can be a piece of hookup wire (insulated) with an alligator clip affixed to the free end. This, in turn, can be clipped to any convenient large metal object which will serve as the antenna. Also, see Table 29-1.

Table 29-1. Parts List for Regenerative Radio Receiver.

Item No.	Description
B1	9-volt battery.
C1	365-pf variable capacitor.
C2	.022-mfd capacitor.
C3	.0047-mfd capacitor.
C4	51-pf capacitor.
L1	Transistor antenna coil (Lafayette MS-166 or equiv.).
L2	10-15 turns of hookup wire over the middle of L1.
Q1	2N366 transistor.
R1	2-meg potentiometer with spst switch.
	Earphone.

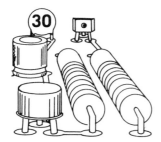

Headset Booster

You will note that a good number of projects in this book call for the use of 1K to 2K-ohm impedance magnetic headphones as the output. In some cases, however, you may want more volume than is supplied in this fashion. Take a peek at the extremely simple circuit that accompanies this project. See Fig. 30-1 and Table 30-1.

With just a handful of parts, you can assemble a powerful little headset booster which can be built into a small plastic parts box and equipped with two jacks—one for the earphones and one for the output of the gadget you have just completed (the one you would like amplified). All you do is plug in the headset at one end, and plug in the previous stage at the other end.

You will note than an "impedance preset" control is included in this circuit; this is to permit you to use this headset booster in a wide variety of applications—not only those normally feeding to magnetic headphones. All you do is adjust R1 until the best distortion-free amplified sound is obtained. SW1 should be a separate spst slide switch, not attached to R1, and should be mounted externally.

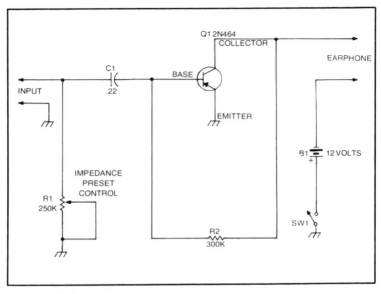

Fig. 30-1. Headset booster schematic.

Table 30-1. Parts List for Headset Booster.

Item No.	Description
B1	12-volt power source (two 6-volt batteries).
C1	.22-mfd capacitor.
Q1	2N464 transistor.
R1	250K potentiometer.
R2	300K resistor.
SW1	Spst switch.
	Earphone.

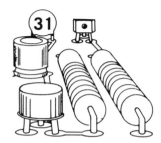

Transistorized Electronic Timer

Here is an interesting experiment in time delay circuitry in addition to being a useful item to have around the workshop. It can be used to switch on and off virtually any electric device you happen to have such as small motors, photographic enlargers, or small electric heaters. The delay time depends on the capacitor you use for C1. With the 25-mfd electrolytic we used, you can adjust for a delay of 1 to 12 seconds; with a 100-mfd electrolytic you can expect anywhere from 5 to 50 seconds.

Operation is simple: Turn SW1 on and set R3 for the desired delay time. The relay will hold down for as long as the time you have dialed on R3, then release, To repeat this procedure, close the reset key and once again close SW1. Construction is fairly simple and straightforward. Make certain that you observe proper polarities of the battery (a 9-volt transistor radio type) and electrolytic capacitor C1. See Fig. 31-1 and Table 31-1.

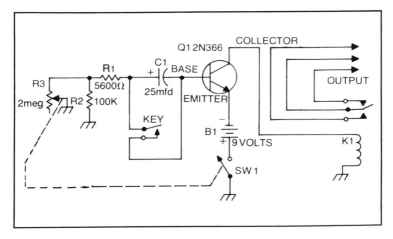

Fig. 31-1. Electronic timer circuit.

Table 31-1. Parts List for Electronic Timer.

Item No.	Description
B1	9-volt battery.
C1	25-mfd, 15-volt electrolytic capacitor.
K1	4000-ohm spdt relay (Advance SO/IC/4000D or equiv.).
Q1	2N366 transistor.
R1	5600-ohm resistor.
R2	100K resistor.
R3	2-meg potentiometer with spst switch.
	Key.

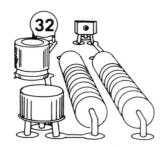

Audio Filter Unit

Here is a unique gadget that you can use in your hi-fi or stereo system, your shortwave listening, your ham receiver, or PA system. It will effectively "null out" annoying noises such as ignition bursts and ac hum without creating any adverse effect on normal listening.

All you do is insert the audio filter unit between the main unit and the speaker or headset. Next, you advance R7 to approximately midway. Adjust control R3 until the noise is effectively nulled down. If it does not altogether disappear at a certain setting, just advance R7 further until it does. With everything complete, adjust R8, the main gain control for the desired speaker or headphone level. To obtain best output without distortion, experiment with the volume control on your amplifier or receiver and control R8.

For most effective results in an audio system, the filter should be inserted between the preamplifier and main amplifier. Bear in mind, however, that the filter must be completely enclosed and carefully wired so as to prevent outside noises from accidentally entering through sheer proximity. Use shielded cabling throughout and keep all exposed leads as short as possible. See Fig. 32-1 and Table 32-1.

Rule-of-thumb: The best results will be obtained when the filter is inserted in an appropriate spot *before* the main amplification takes place. In a receiver, it might be advisable to insert this just after the detector stage.

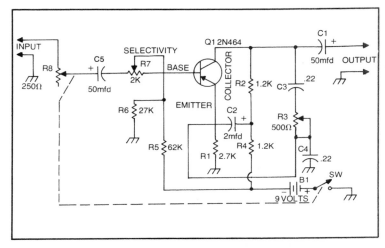

Fig. 32-1. Audio filter circuit.

Table 32-1. Parts List for Audio Filter Unit.

Item No.	Description
B1	9-volt battery.
C1, C5	50-mfd, 15-volt electrolytic capacitors.
C2	2-mfd, 15-volt electrolytic capacitor.
C3, C4	.22-mfd, 75-volt capacitors.
Q1	2N464 transistor.
R1	2.7K resistor.
R2, R4	1.2K resistors.
R3	500-ohm potentiometer.
R5	62K resistor.
R6	27K resistor.
R7	2K potentiometer.
R8	250-ohm potentiometer with spst switch.

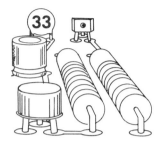

Crazy Kiddie Toy

This unusual gadget is guaranteed to amuse youngsters for hours in addition to providing an unusual method of teaching the fundamentals of music. It consists of a mysterious wooden box containing a wide assortment of switches and a built-in speaker. None of the switches is labeled and the first trick for the child is to locate SW1, the power switch. The other switches, when depressed, produce musical tones, and are sequentially complex, since a different tone can be produced by the same switch depending on which of the preceding switches has also been closed. With an extremely intelligent child, it is possible to figure out a means of playing simple melodies with this gadget. See Fig. 33-1 and Table 33-1.

Nothing is critical in construction, except that the gadget be built to withstand rugged treatment. With youngsters under 10 years of age, it should really be built to take a beating. Just remember to place the switches in such a way as to thoroughly confuse the uninitiated. Incidentally, feel free to throw in whatever kind of switches you have handy: slide switches, toggle switches, microswitches, and even potentiometers with spst contacts.

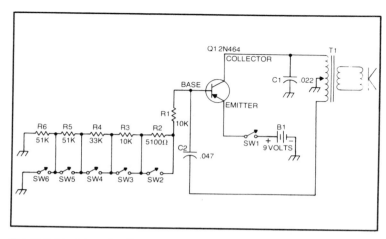

Fig. 33-1. Music box circuit.

Table 33-1. Parts List for Crazy Kiddie Toy.

Item No.	Description
B1	9-volt battery.
C1	.022-mfd capacitor.
C2	.047-mfd capacitor.
Q1	2N464 transistor.
R1, R3	10K resistors.
R2	5100-ohm resistor.
R4	33K resistor.
R5, R6	51K resistors.
SW1, SW2, SW3, SW4, SW5, SW6	Spst switches.
T1	Audio output transformer (Argonne AR-170 or equiv.). Speaker.

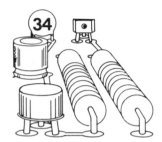

Sensitive Geiger Counter

Would you like to find some uranium in your backyard? You will never know if it is there or not without the aid of a good Geiger counter which, incidentally, can be used to detect radioactivity just about anywhere, including your luminous-dial wristwatch.

Our suggestion is to mount the whole affair in a completely enclosed metal housing, such as an aluminum *Mini-box*. Radioactivity, unlike rf energy, easily penetrates metal, so there is no need to drill holes to let the radioactivity in. If it is nearby, the Geiger counter will pick it up. Attach a small handle to the top of the box, and equip the unit with a dpst power switch accessible from the outside. The only other addition necessary is an earphone jack. See Fig. 34-1 and Table 34-1.

Incidentally, the Geiger tube is an extremely delicate device and should be shock-mounted inside the *Mini-box* with foam rubber taped to the outside of the tube. Make certain everything is well mounted inside to prevent anything from breaking loose if the counter is jarred.

With the earphones in place, you will hear a faint clicking sound. This is normal and is nothing to get excited about. When the counter is brought into the vicinity of a radioactive material, however, this will become increasingly rapid and louder. The faster the clicking, the closer you are to the source of radioactivity.

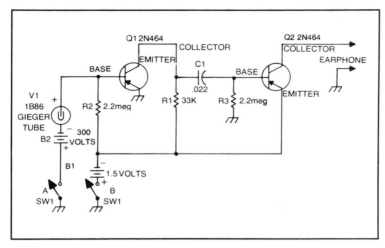

Fig. 34-1. Geiger counter circuit.

Table 34-1. Parts List for Geiger Counter.

Item No.	Description
B1	1.5-volt dry cell.
B2	300-volt battery.
C1	.022-mfd capacitor.
Q1, Q2	2N464 transistors.
R1	33K resistor.
R2, R3	2.2-meg resistors.
SW1	Dpst switch.
V1	1B86 Geiger tube.
	Earphone.

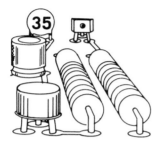

Simplest Audio Amplifier

You may look at the accompanying schematic in utter disbelief, but the plain fact is that with just these four components you can amplify just about any audio signal. It can be used as a phono preamplifier, a microphone booster, a tracer probe, a crystal set amplifier, or almost anything else that requires a bit more volume.

It is so small that it can literally be built *into* an earphone (use a mercury hearing-aid battery), and the whole affair can be put together within an hour, yet it can provide months and months of use without requiring the battery to be replaced.

Construction could hardly be called difficult or critical, just make sure that the positive battery terminal goes directly to ground. Note additionally that although one side of the input goes to ground, the output leads do *not*. This is important if you are to realize optimum performance from your simplest audio amplifier. See Fig. 35-1 and Table 35-1.

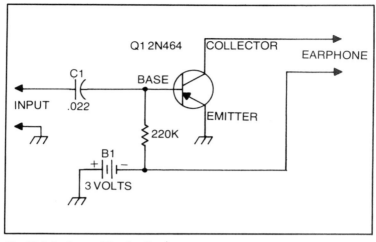

Fig. 35-1. Audio amplifier circuit.

Table 35-1. Parts List for Audio Amplifier.

Item No.	Description
B1	3-volt battery.
C1	.022-mfd capacitor.
Q1	2N464 transistor.
R1	220K resistor.
	Earphone.

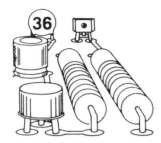

Sun-Powered Code Practice Oscillator

Here is another in our line of "simplest" circuits you can build. While at first glance it may look somewhat complex to the rank beginner, take another look. Most of what you see is simply interconnecting wires. Componentwise, there are only six parts.

The heart of the sun-powered code practice oscillator is, of course, the photocell whch provides power for the transistor. The more light that strikes the face of the photocell, the louder the tone in your headset. No power switch is necessary, since there is no battery to require replacement.

Note the hookup of T1 which, contrary to common usage, is wired in reverse-fashion. Follow the schematic exactly, and you will have no trouble whatever in getting your code oscillator to operate. The key can be any surplus telegraph key, although any kind of switch will trigger the unit into operation. If you do not have a key, a normally, open push-button switch works as well. You can adjust the pitch of the oscillator by simply adjusting R1 to a pleasing level. See Fig. 36-1 and Table 36-1.

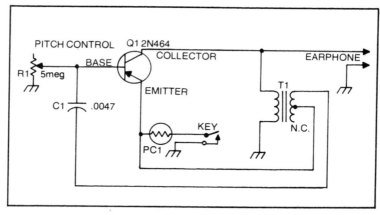

Fig. 36-1. Code practice oscillator circuit.

Table 36-1. Parts List for Code Practice Oscillator.

Item No.	Description
C1	.0047-mfd capacitor.
PC1	Photocell (International Rectifier B2M or equiv.).
Q1	2N464 transistor.
R1	5-meg potentiometer.
T1	Output transformer (Argonne AR-109 or equiv.).
	Key.
	Earphone.

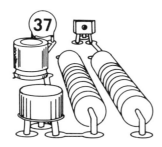

Transistorized Crystal Oscillator

Do you ever wonder if your shortwave receiver is really on the nose calibration-wise? If you have been discouraged trying to find rare dx stations by tuning to the exact frequency specified in a shortwave handbook it is probably due to the fact that your dial is off—perhaps 10 to even 60 kHz.

Build this handy transistor crystal (abbreviated *xtal*) oscillator, however, and you will be hearing those choice foreign stations with no difficulty whatever. The unit is so simple to construct that it can be built right into your receiver chassis. Its operation throws out hundreds of harmonic "signals" every 100 kHz clear across your dial—all the way to 30 MHz. You can use it as a marker generator, by flicking the oscillator on whenever you are in doubt as to frequency, or you can spend an hour or so with the trimmer capacitors of your receiver and recalibrate the whole affair once and for all by checking as you go along with your new 100-kHz generator.

In operation, all you need to do is bring the output lead close to the antenna coil of your receiver and you are in business. The crystal should be an inexpensive 100 kHz marker type, available for approximately $3.00 from most surplus outlets. Tuneup is easy: Monitor on your receiver as you adjust RFC1 and C2 until the oscillator kicks in and signal is clearly heard in the receiver. Mainly, C2 acts as a trimmer, while RFC1 sets the main oscillatory kick-in point. This should be adjusted so that is it just a hair past the triggering point.

Check oscillator performance several times by throwing SW1 on and off. This will tell you ahead of time whether the setting of RFC1 is correct. What should happen is that the 100-kHz signals should come on every time SW1 is turned on. See Fig. 37-1 and Table 37-1.

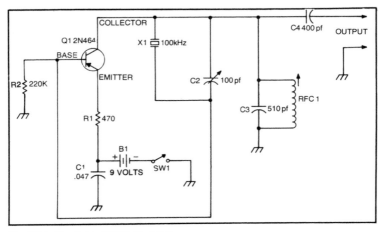

Fig. 37-1. Crystal oscillator diagram.

Table 37-1. Parts List for Crystal Oscillator.

Item No.	Description
B1	9-volt battery.
C1	.047-mfd capacitor.
C2	100-pf variable capacitor.
C3	510-pf capacitor.
C4	400-pf capacitor.
Q1	2N464 transistor.
R1	470-ohm resistor.
R2	220K resistor.
RFC1	35-mh adjustable rf choke (Superex V-25 or equiv.).
SW1	Spst switch.
X1	100-kHz surplus marker crystal.

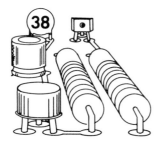

Inexpensive Direct-Coupled Amplifier

Are you looking for a powerful, yet inexpensive-to-build audio amplifier? All you need are two transistors, two capacitors, three resistors, a switch, and 6-volts worth of flashlight batteries and you are in business. This unique little set is highly efficient, draws very little current from the battery pack, yet provides clear distortion-free audio for a pair of ordinary magnetic headphones. Best of all, the unit is entirely portable and so small that it can be built into a standard-sized plastic electronic parts box.

Depending on its use, wiring can be simple or quite critical. If, for example, you are going to simply amplify the output of a crystal or one-transistor radio receiver project, you will have no difficulty whatsoever. If, however, you are going to feed the input leads directly to the ceramic or crystal cartridge of a record player, you will need to take certain precautions to prevent the unwanted inducement of extraneous noise and ac hum. Best bet in any case is to construct the amplifier in a *Mini-box* and use shielded mike cabling on all leads carrying audio. See Fig. 38-1 and Table 38-1.

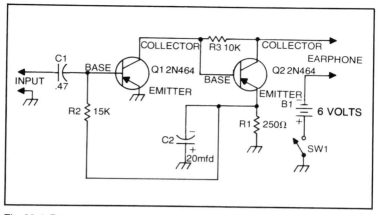

Fig. 38-1. Direct-coupled amplifier circuit.

Table 38-1. Parts List for Direct-Coupled Amplifier.

Item No.	Description
B1	6-volt power source (four 1.5-volt dry cells).
C1	.47-mfd capacitor.
C2	20-mfd, 10-volt electrolytic capacitor.
Q1, Q2	2N464 transistors.
R1	250-ohm resistor.
R2	15K resistor.
R3	10K resistor.
SW1	Spst switch.
	Earphone.

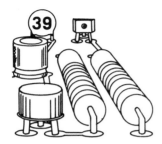

House-Wire CW Set

Are you interested in a private communications system over which you can send secret ciphered messages without requiring any more than just plugging into the nearest ac wall outlet? This project consists of a unique transmitter which transmits AM broadcast-band signals to any point in your house through the 117-volt ac house wiring. In many cases, it will also send your secret signals to a friend's house, or even to the top floor in your apartment building. All you need to do to hear the secret messages is to plug a regular AM radio into any convenient outlet and tune to the transmit frequency. The coded signals are heard loud and clear.

The circuit shown in the accompanying diagram consists of a code practice oscillator (so that you can hear the tones exactly as they are heard by your remote eavesdropper) feeding into a small speaker and an AM broadcast band transmitter. The antenna is the ac electric house wiring. The oscillator, in addition to providing a pleasing monitor note, also modulates the transmitted carrier. Hence, a perfectly clear signal is received on the remote radio.

To operate the system, depress the key, instantly firing up the transmitter and oscillator, and adjust the slug of L2 to any unused frequency in the 550-1650 kHz AM broadcast band. For a true two-way system, build two CW sets and have each communicator equipped with a table radio tuned to the secret broadcast frequency. See Fig. 39-1 and Table 39-1.

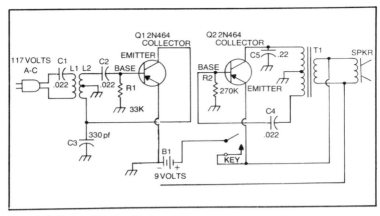

Fig. 39-1. Transmitter circuit.

Table 39-1. Parts List for CW Oscillator-Transmitter.

Item No.	Description
B1	9-volt battery.
C1, C2, C4	.022-mfd capacitors.
C3	330-pf capacitor.
C5	.22-mfd capacitor.
L1	11 turns of No. 22 enameled wire, wound on end of L2 furthest from connection to Q1.
L2	Tapped variable loopstick (Superex VLT-240 or equiv.).
Q1, Q2	2N464 transistors.
R1	33K resistor.
R2	270K resistor.
T1	Audio output transformer (Argonne AR-120 or equiv.).
	Key.
	Speaker.

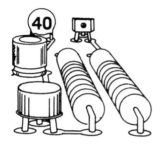

CW Monitor

While it is all well and good to say that you do not need to hear what your keying sounds like, frequently the poor chap at the other end would say something quite different. In fact, if most hams could hear the poor quality of code they send over the ham bands, they would be astonished, but monitors are expensive and many CW operators do not regard them as necessities.

Our advice, however, particularly to novices, is to build the easy-to-construct gadget illustrated in the accompanying schematic. It requires no battery, will never burn out, yet will provide a pleasant tone in any pm speaker. So small that it can be built right into the transmitter, it is powered entirely by induction: it steals rf current from the antenna line and uses it to drive the transistor and speaker.

Nothing is critical about construction. The antenna loop L1 is merely a 2-turn link wound over your antenna wire (or lead feeding internally to the antenna receptacle on your transmitter). If a strong tone is not produced, try a 5-turn loop. Generally, the better the coupling, the louder the tone. Novices using long-wire antennas have an easy job with this circuit. All they have to do is wind 5 turns or so of ordinary hookup wire around any part of the antenna wire and run both ends back to the CW monitor. See Fig. 40 and Table 40.

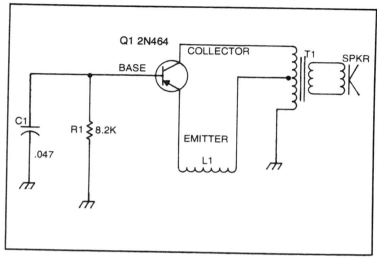

Fig. 40-1. CW monitor circuit.

Table 40-1. Parts List for CW Monitor.

Item No.	Description
C1	.047-mfd capacitor.
L1	2-turn link (see text).
Q1	2N464 transistor.
R1	8.2K resistor.
T1	Audio output transformer (Argonne AR-170 or equiv.).
	Speaker.

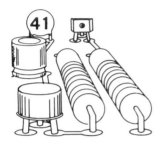

Hi-Fi Audio Mixer

Did you ever want to combine two inputs into one *before* they reached your amplifier or tape recorder? You can do it swiftly and inexpensively with this little hi-fi audio mixer.

The unit can be used for mixing programmed material (such as music from an FM tuner) and live material (such as a vocalist). You can achieve any kind of dominance you wish simply by adjusting R5 and R6. For example, you can insert the FM programming as background music and have your live voice dominate the recording; or you can work a combination of effects simply by adjusting these controls.

Construction is critical in that the unit must be completely enclosed in a metallic housing (such as an aluminum *Mini-box*) and all leads must be as short as possible. Use shielded mike cabling on all wires carrying audio. The only external controls needed are the two input (mixing) potentiometers and the slide switch SW1. Suitable phono jacks should be used for both inputs and the output. See Fig. 41 and Table 41.

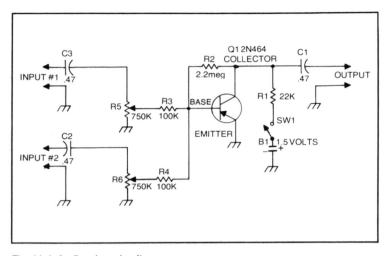

Fig. 41-1. Audio mixer circuit.

Table 41-1. Parts List for Audio Mixer.

Item No.	Description
B1	1.5-volt dry cell.
C1, C2, C3	.47-mfd capacitors.
Q1	2N464 transistor.
R1	22K resistor.
R2	2.2-meg resistor.
R3, R4	100K resistors.
R5, R6	750K potentiometers.
SW1	Spst switch.

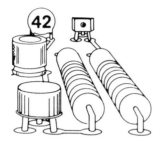

Push-Pull Receiver

Here is a highly unusual transistor-semiconductor project that will prove particularly interesting to the experimenter who has already constructed several different types of AM radio receivers. It makes maximum use of just a handful of components to provide a professional, sensitive radio receiver that is selective and powerful. In fact, you will have enough audio output to power a small pm speaker.

Note the bridge-like formation of four crystal diodes. This acts as the detector complex, leaving the transistors free to perform as full-time push-pull amplifiers. Coil L1 is an ordinary vari-loopstick (untapped) antenna coil which is adjusted for calibration purposes only. When the plates of capacitor C3 are fully meshed, you adjust L1 so that the receiver is set at approximately 550 kHz on the broadcast band. After L1 is set, C3 is your tuning control.

The antenna can be a short length of ordinary insulated hookup wire with an alligator clip affixed to the free end. The clip can be attached to any convenient metallic (ungrounded) object which will serve as a giant receiving antenna. See Fig. 42-1 and Table 42-1.

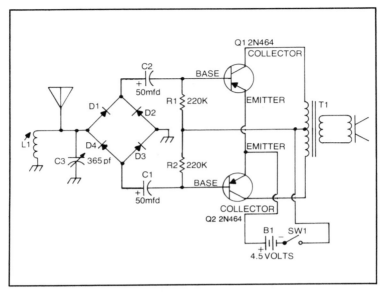

Fig. 42-1. AM receiver schematic.

Table 42-1. Parts List for Push-Pull Receiver.

Item No.	Description
B1	4.5-volt power source (three 1.5-volt dry cells).
C1, C2	50-mfd, 15-volt electrolytic capacitors.
C3	365-pf variable capacitor.
D1, D2, D3, D4	1N34A diodes.
L1	Variable loopstick antenna coil (Superex VLT-240 or equiv.).
Q1, Q2	2N464 transistors.
R1, R2	220K resistors.
SW1	Spst switch.
T1	Transistor output transformer (Argonne AR-170 or equiv.).
	Speaker.

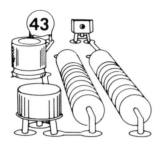

Personal Metronome

Are you a piano player or trying to master the organ? Well if you were not born with a feel for music, it is no easy job to become an accomplished musician; in fact, you are going to need all the help you can get. Without bothering anyone with annoying loud clack-clacks, you can still enjoy all the benefits of a professional metronome with the gadget shown schematically in the accompanying diagram. All you do is place the headset over your ear, dial the "beat" you want, and start playing. As long as you synchronize your left-hand accompaniment with the metronome sound in the earphones, you will sound like you have been playing professionally for years.

Actually so compact that you can build it into an empty hard-pack cigarette pack, this little metronome produces adequate volume from just two transistors, a few stray parts, and a standard 9-volt transistor radio battery. To operate, just turn on SW1 and dial the proper beat by adjusting R1. Construction is not critical. If you like, you can substitute a .47-mfd capacitor for C1 and C2. See Fig. 43-1 and Table 43-1.

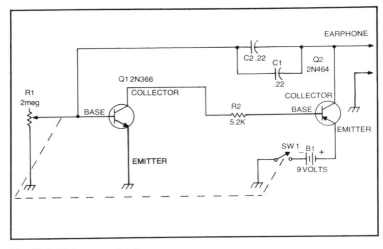

Fig. 43-1. Metronome circuit.

Table 43-1. Parts List for Personal Metronome.

Item No.	Description
B1	9-volt battery.
C1, C2	.22-mfd capacitors.
Q1	2N366 transistor.
Q2	2N464 transistor.
R1	2-meg potentiometer with spst switch.
R2	5.2K resistor.
	Earphone.

95

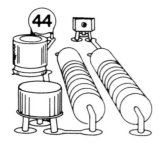

Two-Transistor Sensitive Light Relay

Are you interested in a gadget that will turn appliances on and off in a jiffy everytime a light beam hits the photocell sensor? This one is just such a device. In fact, even the low light from a tiny pocket pen light will trigger this unit into operation from a distance of 25 feet.

To be honest, the circuit has been "stolen" from a professional store-door alarm. You have seen the kind; this is the device in which an alarm bell rings every time a customer interrupts the light beam across the foot of the doorway. You can work your alarm or off/on switch in any fashion you choose, simply by making appropriate connections to the relay contacts.

Sensitivity can be adjusted or preset by simply adjusting the setting of potentiometer R3, which also serves as the off-on switch for the light relay. The photocell should be placed outside the housing in a position where it will receive the light beam you send to it. Incidentally, this makes a great daylight alarm. See Fig. 44-1 and Table 44-1.

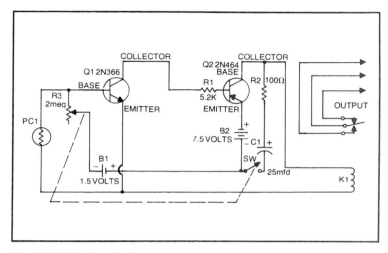

Fig. 44-1. Light relay circuit.

Table 44-1. Parts List for Sensitive Light Relay.

Item No.	Description
B1	1.5-volt dry cell.
B2	7.5-volt battery.
C1	25-mfd, 15-volt electrolytic capacitor.
K1	4000-ohm spdt relay (Advance SO/IC/4000D or equiv.).
PC1	Photocell (International Rectifier B2M or equiv.).
Q1	2N366 transistor.
Q2	2N464 transistor.
R1	5.2K resistor.
R2	100-ohm resistor.
R3	2-meg potentiometer with spst switch.

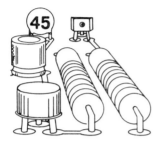

Blinker

We have all seen those interesting flashing neon signs, but the big problem in do-it-yourself projects has been to obtain the proper voltage necessary to "fire" the bulb into illumination. With the unusual device shown in the accompanying schematic you can get I1 not only to light but also to flash intermittently—and all from a 6-volt battery.

The heart of the unit is, of course, the universal output transformer which tends to build up the six volts to a higher level in order to operate the bulb. This transformer is not critical, and can be just about any universal type you have handy. The rate of flashing of the neon bulb is largely determined by where you set potentiometer R1. You will note that there is no power switch in this circuit; the set draws so little current that it will flash almost forever from just four size-D flashlight cells. See Fig. 45-1 and Table 45-1.

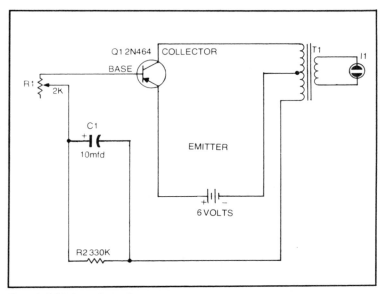

Fig. 45-1. Blinker schematic.

Table 45-1. Parts List for Light Blinker.

Item No.	Description
B1	6-volt battery.
C1	10-mfd, 10-volt electrolytic capacitor.
I1	NE-2H bulb.
Q1	2N464 transistor.
R1	2K potentiometer.
R2	330K resistor.
T1	Universal output transformer.

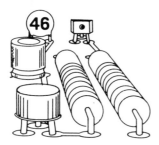

Square-Wave Audio Generator

While audio tone oscillators are great for ordinary purposes, you are going to need a dependable square-wave generator if you are interested in doing a truly professional troubleshooting job on a piece of defunct electronic equipment. Such is the device shown in the accompanying diagram.

This unit is powered by 117-volt ac household lines, so you will want to make certain that there is no chance of an electrical short in your circuit. Other than this, no great hazards are present. You can build the unit in a small aluminum box, taking care to use rubber grommets wherever leads pass in or out of the enclosure. See Fig. 46-1 and Table 46-1.

Operation is simple. The output is a steady 1 volt of solid 60-Hz (cycle) signal. You can adjust this output down by simply changing the setting of R1. You can check out your generator by hooking it up directly to an oscilloscope and checking the square-wave pattern on the screen. Incidentally, if you want to inject an outside sine wave (from an external generator), you can connect the common external generator ground lead to your square-wave generator chassis and the "hot" lead to the junction of R4 and the secondary of T1. If you do this, remove the end of R4 from T1 so that the 60-Hz signal is temporarily disconnected. In this manner you can enjoy the fullest versatility of this test instrument—at a minimum expense.

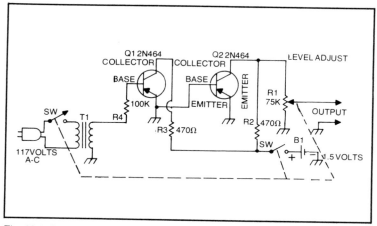

Fig. 46-1. Square-wave generator circuit.

Table 46-1. Parts List for Square-Wave Generator.

Item No.	Description
B1	1.5-volt dry cell.
Q1, Q2	2N464 transistors.
R1	75K potentiometer with dpst switch.
R2, R3	470-ohm resistors.
R4	100K resistor.
T1	Power transformer (Stancor P-6465 or equiv.).

Automatic Sound Switch

Here is an unusual project for the beginner in transistor experimentation who really wants to come up with something useful. Although this has numerous applications, we designed it specifically as a garage door opener that works by sound only. As your car approaches the garage, you sound the horn. The switch detects this blast and signals the relay to start the garage door opener circuit. You drive into the garage; as you leave, the garage door closes—automatically.

Here is how it works: Adjust potentiometer R2 to "hear" only the sound you want to trigger the circuit. Next, adjust R3 for the approximate "time-on" period, which can be anywhere from 1 to 20 seconds, depending upon where you set the control. From here, everything is automatic. Nothing happens until the speaker hears the tone or voice signal it requires. Hookup can be in any fashion you want; just connect to the contacts you wish to use on K1. See Fig. 47-1 and Table 47-1.

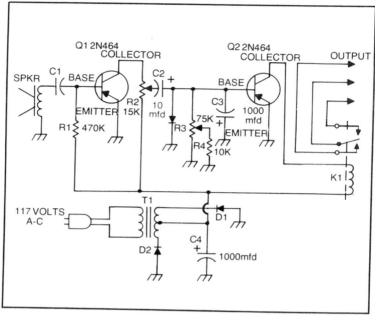

Fig. 47-1. Automatic sound circuit.

Table 47-1. Parts List for Automatic Sound Switch.

Item No.	Description
C1	50-mfd, 25-volt electrolytic capacitor.
C2	10-mfd, 25-volt electrolytic capacitor.
C3, C4	1000-mfd, 25-volt electrolytic capacitors.
D1, D2, D3	1N38B diodes.
K1	4000-ohm spdt relay (Advance SO/IC/4000D type or equiv.).
Q1, Q2	2N464 transistors.
R1	470K resistor.
R2	15K potentiometer.
R3	75K potentiometer.
R4	10K resistor.
T1	Filament transformer, 12.6-volt ac type.

Audio Frequency Meter

Have you ever wanted to accurately measure the frequency of a particular audio tone to find out, for example, whether it is 1000 or 500 Hz? With the portable audio frequency meter shown in the accompanying schematic, you can measure any specific value from 20 to 1000 Hz. Plug an audio signal source into the input and read the output on the 0-1 ma meter. If, say, you get a reading of .62 ma, you are looking at an accurate measurement of 620 hertz.

The entire affair can be built into a small aluminum meter box. Only one external control is required: the power on-off switch, SW1, which can be a simple slide switch. The 9-volt transistor radio battery is contained within the meter housing. See Fig. 48-1 and Table 48-1.

Incidentally, this circuit responds best to well-amplified sources. Somewhere between 1 and 2 volts input is the requirement we found, although this can easily be remedied by simply inserting a preamplifier in the input line if you intend to do a lot of measuring of minute af signals.

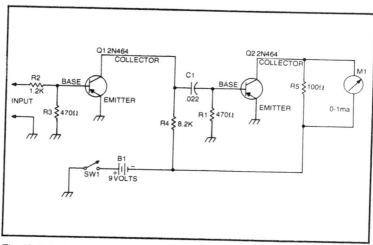

Fig. 48-1. Audio frequency meter schematic.

Table 48-1. Parts List for Audio Frequency Meter.

Item No.	Description
B1	9-volt battery.
C1	.022-mfd capacitor.
M1	0-1 dc milliammeter.
Q1, Q2	2N464 transistors.
R1, R3	470-ohm resistors.
R2	1.2K resistor.
R4	8.2K resistor.
R5	100-ohm resistor.
SW1	Spst switch.

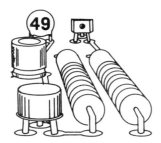

One-Transistor Light Relay

Here is an interesting automatic relay switcher which requires only one transistor, yet is capable of converting light beams into electrical energy every time without fail. It can serve a multitude of applications, including intrusion alarms, morning awakeners, and target games for the youngsters. All that is necessary is that you hook up whatever circuit you want to the relay contacts on K1. The inexpensive light relay will do the rest. It can be designed to switch on an alarm only when the light beam is interrupted, or the reverse—to signal when a light hits the photocell.

Construction is quite simple and straightforward. An aluminum housing makes a good enclosure, but this is not critical. A plastic parts box will do just as well. Watch battery and electrolytic capacitor polarities, and you should experience no trouble. The photocell can be mounted atop your light relay box, or wired remotely as far as 50 feet away without degrading performance. Just make certain you use good nonresistive copper interconnecting wires. See Fig. 49-1 and Table 49-1.

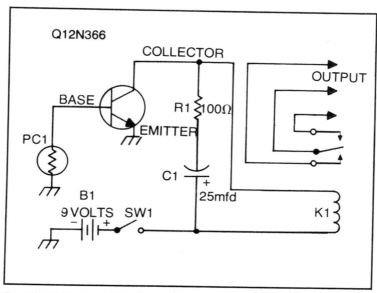

Fig. 49-1. Light relay schematic.

Table 49-1. Parts List for Light Relay.

Item No.	Description
B1	9-volt battery.
C1	25-mfd, 15-volt electrolytic capacitor.
K1	4000-ohm spdt relay (Advance SO/IC/4000D or equiv.).
PC1	Photocell (International Rectifier B2M or equiv.).
Q1	2N366 transistor.
R1	100-ohm resistor.
SW1	Spst switch.

Index

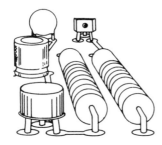

Index

A

AC drive controller	50
circuit	51
parts list	51
Adapter, headset-to-speaker	24
Alarm, moisture/rain	56
All-purpose signal generator	18
Amplifier, audio	78
direct-coupled	84
echo-chamber	20
AM radio, circuit	27
one-transistor	16
AM receiver, high-power	42
parts list	43
schematic	43, 93
Audio amplifier	78
circuit	79
parts list	79
Audio filter unit	72
circuit	73
parts list	73
Audio frequency meter	104
parts list	105
schematic	105
Audio generator, square-wave	100
Audio mixer, circuit	91
hi-fi	90
parts list	91
Audio preamplifier, circuit	53
parts list	53
2-transistor	52
Auto light reminder	54
circuit	55
parts list	55
Automatic AC drive controller	50
Automatic auto light reminder	54
Automatic safety flasher	22
Automatic sound switch	102
circuit	103
parts list	103

B

Battery eliminator	64
parts list	65
schematic	65
Batteryless transistor receiver	26
parts list	27
Blinker	98
parts list	99
schematic	99

C

CB field	
strength meter	34
Code practice monitor	48
parts list	49
Code practice oscillator	14, 49
circuit	15, 81
parts list	15, 81
sun-powered	80
Communications system,	
light beam	58
Controller, auto-	
matic AC driver	50
Crystal oscillator	82
calibrator diagram	83

parts list	83
CW monitor	88
circuit	89
parts list	89
CW oscillator-transmitter, parts list	98
CW radio transmitter	40
CW set, house-wire	86

D

Direct-coupled amplifier	84
circuit	85
parts list	85
Dynamic mike	36
circuit	37
parts list	37

E

Eavesdropper, parts list	47
schematic	47
supersonic	46
Echo-chamber amplifier	20
circuit	21
parts list	21
Electronic moisture/rain alarm	56
Electronic music maker	30
circuit	31
parts list	31
Electronic timer, circuit	71
parts list	71
transistorized	70
Eliminator, battery	64

F

Field strength meter, CB	34
circuit	35
parts list	35
Filter unit, audio	72
Flasher, safety	22
FM radio	32
circuit	33
parts list	33
Frequency meter, audio	104

G

Geiger counter	76
circuit	77
parts list	77
Generator, audio	100
signal	18

H

Hand motion	
music maker	12
parts list	13

Handy signal tracer	38
Headset booster	68
parts list	69
schematic	69
Headset-to-speaker adapter	24
parts list	25
schematic	25
Hi-fi audio mixer	90
High-power AM receiver	42
House-wire CW set	86

K

Kiddie toy	74
parts list	75

L

Light beam	
communications system	58
circuit	59
parts list	59
Light relay, circuit	97
one-transistor	106
parts list	97, 107
schematic	107
two-transistor	96
Light reminder, automatic auto	54

M

Magic meter sounder	28
Meter sounder	28
circuit	29
parts list	29
Metronome, personal	94
Microphone	36
Miniature FM radio	32
Moisture/rain	
alarm, circuit	57
electronic	56
parts list	57
Monitor, code practice	48
CW	88
Music box, circuit	75
Music maker, electronic	30
hand motion	12

O

One-transistor	
AM radio	16
circuit	17
parts list	17
One-transistor light relay	106
Oscillator, code practice	14, 80

sound	13
transistorized crystal	82

P

Personal metronome	94
circuit	95
parts list	95
Powerful code	
practice monitor	48
Preamplifier,	
2-transistor audio	52
Push-pull receiver	92
parts list	93

R

Radio, FM	32
Radio receiver,	
sunlight	60
Radio schematic	61
Radio transmitter,	
circuit	41
legal CW	40
parts list	41
Receiver, AM	42
PUSH PULL	92
radio	60
transistor	26
VHF	44
Regenerative radio, circuit	67
parts list	67
Regenerative radio	
set, beginner's	66
Room metronome, circuit	63
parts list	63
two-transistor	62

S

Safety flasher, automatic	22
circuit	23
parts list	23
Signal generator,	
all-purpose	18

parts list	19
schematic	19
Signal tracer	38
circuit	39
parts list	39
Sound oscillator,	
circuit	13
Sound switch, automatic	102
Sunlight radio receiver	60
parts list	61
Sun-powered code	
practice oscillator	80
Supersonic eavesdropper	46
Square-wave	
audio generator	100
circuit	101
parts list	101

T

Timer, electronic	70
Tracer, signal	38
Transistorized	
crystal oscillator	82
Transistorized	
electronic timer	70
Transistor receiver,	
batteryless	26
Transmitter, radio	40
Two-transistor	
audio preamplifier	52
room metronome	62
sensitive light relay	96

V

VHF receiver, circuit	45
80 to 150 MHz	44
parts list	45

W

Wireless home broadcaster	10
circuit	11
parts list	11